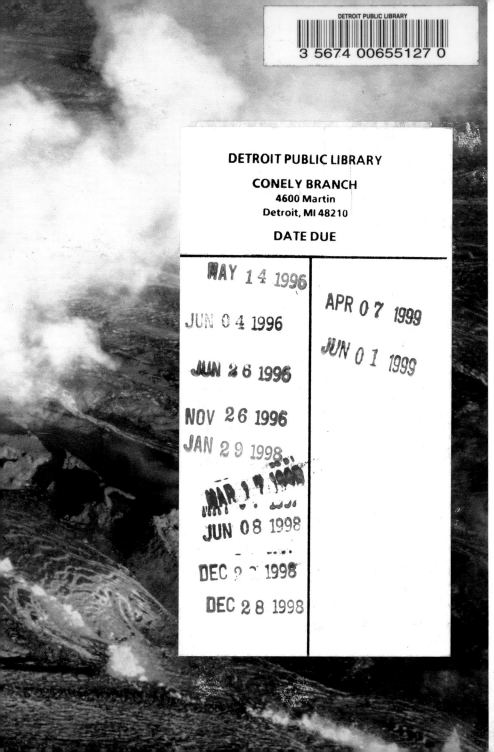

W9-CPL-511

CONTENTS

A volcano is . . . where magma erupts   2
Lava-flows: runny lava   4
Lava-flows: pasty lava   6
Volcanic gas   8
Volcanic explosions   10
Explosive eruptions   12
Pyroclastic-flows   14
Explosions, dust & climate   16
Lava & water   18
Mud-flows   19
Steam-assisted eruptions   20
A volcano is . . . a mountain?   22
A volcano is . . . a hole in the ground   24
Volcanoes around the world   26
Plate tectonics, magma & volcanoes   28
Volcanoes & plates   30
Birth of the Atlantic Ocean   32
An old hot-spot trail   33
Below the volcano   34
Edinburgh volcano   36
Hot springs & geysers   38
Active – extinct – dormant?   40
Hot metal, minerals & water   42
Useful hot water   44
Useful rocks   46
Volcano hazard   48
Volcano watching   50
Coping with an eruption   51
Taupo AD 186   52
Santorini *c.* 1645 BC   53
Vesuvius AD 79   54
Skaftár Fires 1783   55
Krakatau 1883   56
Valley of 10 000 Smokes 1912   57
Mont Pelée 1902   58
Mount St Helens 1980   59

Index & acknowledgements   60

Front cover: fire fountains, Hawaii, 1984.
(Left) a Hawaiian cone cut by a fissure
eruption: Manua Loa, 1984.

1

A VOLCANO IS . . .

Magma is the molten material inside or beneath a volcano. During a volcanic eruption, the magma froths, becoming *lava* and volcanic *gas*.

The lava explodes if the gas bursts out, hurling blocks of rock out of the crater. Clouds of fragments billow out or flow away.

Lava oozes from the crater and from cracks in the volcano. The lava flows downhill, following valleys, spreading out as it reaches flat land.

1 Volcano erupting flows of lava.

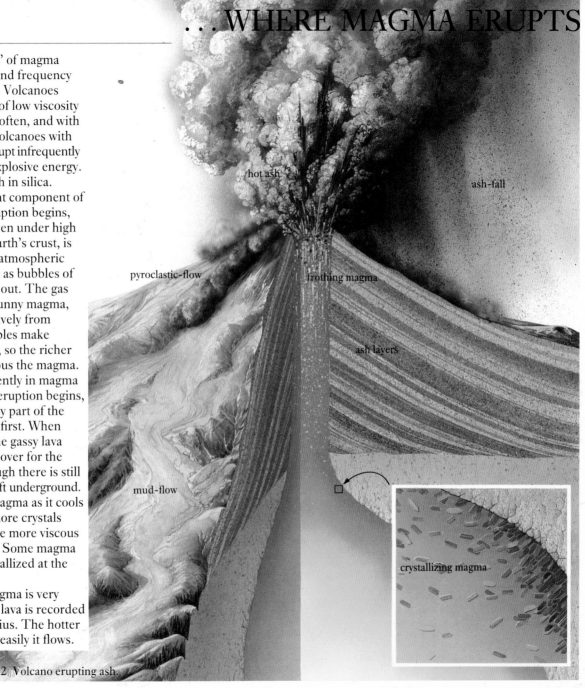

Viscosity or 'stickiness' of magma determines the type and frequency of volcanic eruptions. Volcanoes with runny magma – of low viscosity – tend to erupt more often, and with smaller explosions. Volcanoes with sticky, pasty magma erupt infrequently but with enormous explosive energy. Viscous magma is rich in silica.

Gas is an important component of magma. When an eruption begins, magma, which has been under high pressure inside the earth's crust, is suddenly reduced to atmospheric pressure and it froths as bubbles of volcanic gas separate out. The gas escapes easily from runny magma, but bursts out explosively from viscous magma. Bubbles make magma more viscous, so the richer in gas, the more viscous the magma. Bubbles of gas rise gently in magma underground. As an eruption begins, it is the topmost, gassy part of the magma which erupts first. When there is no more of the gassy lava foam, the eruption is over for the time being, even though there is still some liquid magma left underground.

Crystals grow in magma as it cools underground. The more crystals within the magma, the more viscous the magma becomes. Some magma is almost wholly crystallized at the time of eruption.

Temperature of magma is very variable. The hottest lava is recorded at 1200 degrees Celsius. The hotter the magma the more easily it flows.

hot ash

ash-fall

pyroclastic-flow

frothing magma

ash layers

mud-flow

crystallizing magma

2 Volcano erupting ash.

LAVA-FLOWS: RUNNY LAVA

3 Aa-flow at Mauna Loa, Hawaii, March 1984.

Runny lava may flow from a central crater, a small cone on the flanks of a volcano, or from a crack in the ground. Large eruptions of runny lava tend to build gently-sloping volcanoes. The greatest production of lava is on the ocean floor, where runny lava erupts from fissures in the great system of spreading ridges (pp 28, 30). When runny lava cools it forms a hard, dark rock, basalt. This rock is widespread on the continents as well as on the ocean floor. Basalt lava is erupted at temperatures of up to 1200 degrees C. Dissolved gases are released as bubbles; their force may send the lava up as a 'fire fountain'. As lava cools, bubbles may be trapped as cavities or vesicles. In cooled lava these are often filled with minerals, giving a 'spotty' appearance.

4 Front of an aa-flow, Etna, 1983.

5 'Entrail' pahoehoe-flow, Hawaii, 1987.

When erupting, runny lava behaves much like a river of water. It will flow downhill, following existing valleys, and may form spectacular falls over cliff edges. Long-lasting eruptions may form lava-flows which travel at up to 50km per hour and reach lengths of a hundred kilometres or more. Because it continues to move rapidly, a flow of this length may only be a few metres thick. Runny lava from a fissure may simply form a vast pool, which slowly cools as a thick sheet. The Deccan Plateau of north-west India (p 33) is made of many such sheets.

As a lava-flow cools it becomes stickier and slower-moving. Lava that solidifies with a rough, clinkery surface is known by a Hawaiian name, 'aa' (figs 3, 4). In a thick flow, the lava cools to form solid basalt, which may crack as it contracts to form regular columns. Lava which develops a continuous skin (fig 5) is known by another Hawaiian name, 'pahoehoe'. Its skin can be cool enough to walk on when only a few centimetres thick. If flow continues, the skin becomes wrinkled, resembling coils of rope. In time a thick crust forms; a central tube of flowing lava (figs 6, 7) may empty to form a long cave once the eruption is over. Lava that erupts under-water is rapidly chilled and may break up into tiny glassy fragments or 'pillows'.

Lava-flows rarely cause death, but may destroy agricultural land, roads, bridges and buildings. Attempts have been made to slow threatening flows by cooling them with water hoses, or to divert them with explosives or barriers.

6 Pahoehoe tube, Hawaii.

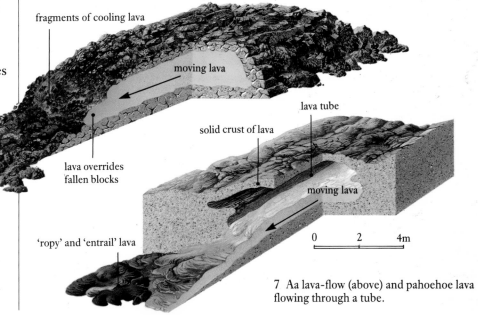

fragments of cooling lava

moving lava

lava overrides fallen blocks

lava tube

solid crust of lava

moving lava

'ropy' and 'entrail' lava

0   2   4m

7 Aa lava-flow (above) and pahoehoe lava flowing through a tube.

LAVA-FLOWS: PASTY LAVA

Lava containing large amounts of silica is very viscous, or pasty. Although runny lava is erupted from volcanoes worldwide, pasty lava is more restricted, mainly to continental edges and strings of islands such as the Caribbean and Japan. Depending on its composition, pasty lava cools to form rocks such as andesite, trachyte, dacite, obsidian and rhyolite. Most pasty lava is erupted explosively, though if the gas content is low, it issues like stiff toffee. Large-volume eruptions can produce very thick lava-flows which move exceedingly slowly (fig 11).

More often, the lava piles up, over and around the vent as a dome (fig 8). The dome grows over a period of weeks or even years, and may reach a height of several hundred metres. In time these domes, which are generally made of rhyolite, dacite or trachyte, become flanked by screes of fragments – 'crumble breccia'. Hot spots in the dome may be marked by spines of pasty lava, squeezed like toothpaste out of a tube. The classic example of a spine is the 300 metre one which formed on Mont Pelée during the 1902–4 eruption (p 58).

Pasty lava which is more mobile will flow down the slope of the volcano and spread out over the surrounding land at a speed of perhaps a few metres each day. The surface of the flow cools to form a thick stony skin, constantly breaking as the underlying lava moves. This happens along horizontal shear planes, so that the lava acts rather like a pack of cards being slid along. The front and sides of the flow become covered with sheared lava blocks (fig 9), which may hide the molten material altogether. The surface of the flow may be thrown into giant ridges (fig. 10), equivalent to the small wrinkles on pahoeoe lava (p 5). Flows such as this may be hundreds of metres thick, but are seldom more than a few kilometres long. Flows of pasty lava damage less property than fast, long, runny flows which can spread over greater areas.

Lava-flows rarely kill, however; all flows normally move slowly enough for people to avoid them. Lava continues to spread across level ground as long as the flow is still fed by continuing eruption. A pasty lava-flow can be very big and thick (fig 11), yet move imperceptibly slowly.

Normally, on eruption, dissolved gases burst their way out of pasty magma with explosive violence (pp 10 to 13) producing clouds of fine volcanic debris called ash. With moderately explosive eruptions, the volcano becomes more steeply conical than those formed by runny lava. The dangers in these eruptions come from the unpredictable nature of the explosions of escaping gas, which may lead to catastrophic pyroclastic-flows (p 14) and from the collapse of weak, oversteepened slopes.

8　Tarumai lava dome, Hokkaido, Japan.

9 The 1783 lava-flow, Asama, Japan.

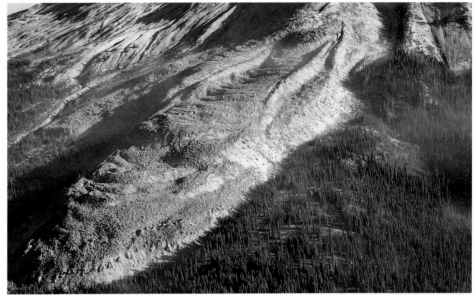

10 Lava-flows 350–450 years old, Mt St Helens south side.

11 Five hundred metre-thick Cerros de Chao lava-flow, N. Chile.

VOLCANIC GAS

12 Bocca Nuova, Etna, 1971.

13 Bocca Nuova by night.

Volcanic gas consists largely of steam and carbon dioxide. Gases were escaping from the molten Earth surface 4.5 billion years ago, forming the atmosphere and oceans. Both free and dissolved, gases are a major ingredient of mineral deposits (p 42).

Carbon dioxide is dense; when erupted, it may collect in valleys as deep pools of gas, suffocating animals. Volcanic gases which contain fluorine, sulphur or chlorine can be very noxious. Such gases were produced by the Skaftár Fires eruption (p 55), with severe effects on vegetation and grazing animals. Sulphur commonly occurs in the form of hydrogen sulphide, giving the 'bad eggs' stench which pervades volcanoes, fumaroles and hot springs. Within the magma these gases are held in solution by pressure inside the Earth. As magma erupts, pressure drops and gases are released. Their manner of escape affects an eruption's explosiveness (pp 10, 11). At Mayon in 1968, gas separation was violent, resulting in exposive eruption (fig 14).

Continuously active, but relatively mildly erupting volcanoes such as Etna provide an opportunity to sample volcanic gases (fig 12). Even so, this is hazardous: the behaviour of volcanoes is unpredictable. Protective suits are worn against temperatures of around 1000 degrees C. A night view (fig 13) reveals the high gas temperature. Several samples are collected taking care to prevent contamination. Useful data can also be obtained by extraction of gases trapped within lava after it has solidified, and by analysis of samples from hot springs.

14 Explosive eruption with pyroclastic-flows, Mayon, Philippines, 1968.

VOLCANIC EXPLOSIONS

Many volcanic eruptions are explosive. Small explosions can be the size of quarry blasts. In historic times, the largest volcanic explosions have been far greater than the largest bombs ever detonated on Earth, but even these are dwarfed by some prehistoric eruptions.

Modern volcanic explosions can be compared with explosions which took place long ago. Even though there were no witnesses, what might have happened can still be worked out by observing the solidified lava and ash. By comparing these with modern day eruptions which were witnessed, a story can be pieced together for their eruption.

Although each volcano is different, explosions can be classified into various types, some called after a representative volcano or notable eruption.

Hawaiian
Volcanic gas escaping from hot and runny basalt lava causes it to spout from a crack or hole. As the cooling lava flies through the air, it is broken up by the escaping gas. Small fragments blow away on the wind, but most are large and settle near the fountain as a cone.

Phreatoplinian & Surtseyan
Eruption into shallow water, the sea or a lake, shatters lava into tiny particles, as water is suddenly converted into steam. The eruption can be a pulsating series of explosions (left) or a continuous roar as water floods in and mixes as a hot lava slurry. Sudden steam expansion powers the outburst of either pulverized lava or a mixture of lava and rock fragments. Phreatoplinian explosions (right) are the very largest of these watery eruptions, and none has happened in historic times.

Strombolian
Frequent explosions burst from a thinly crusted-over lava pool in a crater. When a large bubble of gas rises quickly in runny magma, it bursts the solidified lava skin and explodes from the crater. Fragments of cooling lava from the bursting bubble range from dust-sized to blocks over a metre across. These are thrown out around the crater. The reddish-brown coloured fragments build up a cone, whose slopes collapse whenever they grow too steep. Explosion types are graded from Hawaiian to Strombolian and Plinian, with no distinct boundaries.

Sub-Plinian
Caused by gas foaming from magma deep in the volcano feed-pipe. These are somewhat smaller than Plinian explosions. The fragmented lava and gas is blasted to heights of up to 30 kilometres by the explosion. The explosion column may collapse back around the vent, to travel along the ground as fast-moving pyroclastic-flows, at 100 kilometres an hour or more.

VOLCANIC EXPLOSIONS

Plinian & Ultraplinian

A high velocity explosion (below) shoots a huge column of pumice fragments up to 50 km into the upper atmosphere. This happens when gas foams up in pasty magma below a volcano. Particles settle out hundreds of kilometres around. Nearby towns may be buried within hours under several metres of pumice fragments. It may be several hundred years between one Plinian explosion and the next at one volcano.

No Ultraplinian events have taken place in historical times. Lava fragments are lofted into an explosion column more than 50 kilometres high, and small particles settle over areas the size of a whole continent.

Vulcanian

Cannon-like explosions happen at unpredictable intervals, minutes to several hours apart. An explosion (below) happens when enough gas collects under a plug of lava or rocks to blow it out, maybe even at supersonic speed. The largest fragments are thrown many tens of metres from the crater.

EXPLOSIVE ERUPTIONS

15 Ash-covered terraced fields near Galunggung in eruption, Java, August 1982.

During eruption of magma, gas may escape explosively to produce fragments, which are scattered by the force of the explosion. In mild, Hawaiian type fire-fountain explosions (fig 16), most of the fragments fall nearby. Smaller particles are blown away on the wind. Hawaiian lava is very runny; drops of liquid flying through the air are stretched out into thin threads of volcanic glass. These blow away and collect as 'Pelés hair', Pelé being the Hawaiian fire-goddess. Around small, ashy explosions, fragments fall near the vent and build up a cone. Sloping layers (fig 17) may reveal stages in explosive activity. Larger explosions involve pasty, gas-rich magma, or runny magma and water (p 20) and often produce large quantities of finely pulverized rock and lava particles. This is called 'ash' because it looks like ash from a fire, even though it has not been burned. When an explosion cloud is lofted and spreads into the atmosphere, ash can cause havoc (pp 16,17), blanketing a large area. A thin coating of ash can be beneficial, adding nutrients to the soil, but any more may be a hazard to life and property (fig 15), collapsing roofs and destroying crops, causing breathing difficulties for animals, preventing machinery and transport from working and altering natural drainage. Fragments from volcanic eruptions – 'tephra', or 'pyroclastics' – also include solid pieces of old lava as well as new lava, crystals and frothed glass ('pumice'). Pea- to walnut-sized fragments are termed 'lapilli'; larger pieces are known as 'blocks' or 'bombs'.

The blast from a large explosion sends up a column of gas and dust in a jet-thrust at speeds of as much as 500 metres per second. The eruption column suddenly expands and slows as it engulfs and heats the surrounding air. For the next few kilometres the column rises like thick, turbulent smoke from a chimney, buoyed up by its convecting heat. Above this zone, where the cloud has become similar in density to the surrounding air and the effect of the wind can take over, an umbrella of dust and gas spreads out. The total height depends on the energy of the explosion – the quantity of erupting magma and rock and the rate the eruption column rushes from the vent. Collapse of the explosion column may create a widespread flow of ash and blocks (p 14).

16 Fire fountains in Hawaii.

17 Ash layers, Las Cañadas, Tenerife.

PYROCLASTIC-FLOWS

Explosive volcanoes erupt fragmented magma, ash and rocks (pyroclastics). Large explosions continue for several hours, erupting many cubic kilometres of fragments, mostly ash and pumice. Initially these are carried upwards as a billowing eruption column. The crater becomes widened by erosion of its walls: this allows more magma to froth, fragment and erupt from the reservoir beneath the volcano. The proportion of fragments in the eruption could increase, and the cloud of pyroclastics and gases cannot mix with enough air to remain buoyant. The eruption column then collapses around the vent, and flows of ash and larger fragments spread out from the eruption centre as hot avalanches which are called pyroclastic-flows (fig 18).

The way in which pyroclastic-flows travel, the area they cover, and the thickness of the pyroclastic blanket, depend on the shape and size of the vent, the volume of fragments and the rate at which they are erupted. Flows in the Valley of Ten Thousand Smokes (p 57) were erupted slowly and fill the valley to a great thickness. During part of an eruption at Taupo, 1800 years ago (p 52), over ten cubic km of magma frothed up and erupted in minutes. Flows spread rapidly in all directions and covered a huge area quite thinly. Yet, in the immensity of the Earth's geological record, the Taupo eruption was merely medium-sized. The largest pumice-rich pyroclastic deposits, known as ignimbrite, would have covered areas greater than that of England, with volumes of up to 3000 cubic km.

Pyroclastic-flows are sometimes preceded by a fast-moving surge, a hot blast of fine-grained ash. Fig 18 shows a surge travelling in front of a pyroclastic-flow. The fine ash layer left by a surge is easily eroded by the first rain that falls after an eruption; such ash is rarely seen. The surge can scorch and fell trees. The pyroclastic-flow itself moves at up to 200 kilometres an hour. It carries unsorted fragments from fine ash to large blocks a metre or so across. If it is hot when it comes to rest, then the ash may be welded into solid rock. Large flows normally contain pumice, solidified pieces of frothed magma. Sometimes the ash is cemented by percolating waters a long time after deposition. In fig 22 pinnacles of ash can be seen: these surround parts of a pyroclastic-flow deposit cemented by minerals from rising hot waters or gases. The softer, uncemented material has been worn away. In trying to work out how the eruption looked, geologists have only fragments of evidence to work with. Fig 19 shows ignimbrites from two eruptions which took place three hundred thousand years apart. The lower layer has an irregular top surface: erosion has carved away swathes of ash, leaving pinnacles. The next eruption filled in the hollows. Over thousands or millions of years, repeated pyroclastic-flow eruptions may build up a level plateau (fig 21). The whole plateau, together with a huge, often lake-filled depression at the eruption site, can be thought of as a 'volcano' even though this is very different from the cone-shape so often depicted in diagrams and photographs (see pp 22–23).

18 Pyroclastic-flows at Augustine, Alaska.

19 Bandelier Tuffs, New Mexico.

21 Plateau formed by the Bandelier Tuffs, San Diego Canyon, New Mexico.

20 Ignimbrite on ash and soil.

22 Pinnacles of cemented ash, La Cueva, New Mexico.

EXPLOSIONS, DUST & CLIMATE

23 Ash from Galunggung, Java.

After a large volcanic explosion, a column of gas and dust rises and then spreads out as a cloud in the atmosphere. The highest columns spread far out, rather than settling as thick deposits immediately around the volcano. Large explosions happen when gas-rich, pasty or 'sticky' magma is erupted (p 13). Apart from large blocks and bombs (fig 24), minute rock particles with glassy fragments, pumice and crystals are erupted, with vapour and liquid droplets including water and sulphuric acid. The cloud from a really large explosion towers into the upper atmosphere; stratosphere winds spread it around the Earth. Most dust particles settle out in the months following eruption. However, droplets of sulphuric acid can remain aloft for several years.

So-called 'dust-veils' can have a spectacular effect upon the atmosphere, producing the most beautiful sunsets and causing strange optical effects, even a blue moon. Our climate may be affected by the haze of sulphuric acid droplets. Sulphur occurs in magma and is frequently found around fumaroles. It causes much of the powerful smell around volcanic eruptions and hot springs. Volcanic explosions send sulphur into the atmosphere in the form of gases such as sulphur dioxide and hydrogen sulphide. The action of sunlight and water vapour causes these to re-form, over many weeks to months, into an aerosol of sulphuric acid droplets. These droplets absorb the sun's radiation and scatter it back into space, reducing the average global temperature. The 1982 eruption of El Chichon, Mexico, spread a veil of sulphuric acid droplets around the world in less than a month. This was a particularly sulphur-rich eruption. With around 20 million tonnes of sulphuric acid in the stratosphere one month afterwards it caused a slight but rapid change in northern hemisphere temperatures.

Details of the effects of eruptions upon our climate are difficult to interpret. A record is being constructed of the effects of climate on yearly tree rings, which go back several thousand years. Effects of many large eruptions show up in frost-damaged rings from trees in certain areas. Relatively small but sulphur-rich eruptions in the past have had a great influence on global temperature.

24 Bread crust bomb, Asama, Japan.

Great eruptions leave widespread dustings of volcanic ash in mud, peat and polar ice layers. Using these, a time record of climate changes and disruptions is being built up. Together with other studies of climate, life and rocks, these records are likely to help with prediction of further climate changes and maybe even with the prevention of some of their damaging effects. However, there is much still to be studied and discovered.

Ash layers from numerous eruptions, including many in the remote past, provide a picture of events which helps to piece together the Earth's historical time scale. Since these layers can be dated, a time scale of many hundreds of millions of years is being constructed in ever greater detail.

26 Evening view of Mt St Helens explosion cloud, July 1980.

25 Cinder cones on Etna's south slope.

27 Sunrise colour after the eruption of El Chichon, Mexico, October 1982.

LAVA & WATER

When molten rock meets water, steam is instantly formed. The effects are very variable, however. Where runny lava erupts on land and then meets the sea at the coastline, the lava surface is chilled: the shock breaks much of the lava into tiny fragments of black glass to make beaches of black sand.

On the sea-bed, a new lava surface may be crackled and broken into small, glassy granules as it cools. If runny lava erupts in small amounts, a glassy 'rind' hardens around each extrusion. The rind splits (fig 29) to allow further extrusion, and 'pillow lava' (fig 28) results; the interior hardens as finely crystalline rock. Faster eruption can mix runny lava and water efficiently enough for steam to explode violently; the shallower the sea, the more forcibly steam expands.

Commonly, more voluminous floods of runny lava spread across the sea-bed: large tracts of deep ocean floor are made up of such flows.

Eruptions are affected by water from sources other than the sea, for instance, underground water in rocks (p 20), crater lakes and icecaps. When seeking evidence for volcanoes in ancient rocks, the study of both pillow lavas and the altered remains of sand-like glassy granules (known as hyaloclastite: 'broken-glass rock') can help in identifying the watery environment and the style of eruption. However, there has to be careful distinction between water-generated granules and airfall ash, and between pillows and the lobes of smooth lava known as pahoehoe (p 5).

28 Pillow lavas, Canary Islands.

29 Pillows forming, Kilauea, Hawaii.

30 Black sand and boulders, Kalapana Beach, Hawaii.

Explosive eruptions leave a volcano mantled in a layer of ash. Addition of even a small amount of water mobilizes the ash, giving it the consistency of wet cement. In this state it may turn into a potentially lethal mud-flow or 'lahar'. The water may be from a number of different sources. For example, at Kelut, Java, in 1966 a crater lake was breached and a flood released; at Ruiz, Colombia, in 1985 ice and snow rapidly melted from upper slopes. Other mud-flows, such as those on the lower slopes of Mayon Volcano, Philippines (p 9), are set off when torrential rain soaks the ash, increasing its weight and reducing cohesion until it is no longer stable. Whatever the cause, the result is much the same. The mud-flow accelerates downslope, to speeds of 100 km per hour. It picks up stones, boulders, tree trunks, remains of buildings and bridges. The flow is narrow in a river valley; otherwise it spreads out as a wide sheet. Flows 100 km long are not uncommon, and ancient flows are known that cover an area of 5000 square kilometres. The deposit left after water has drained away is a tumble of boulders, dust and other debris. Heavy rain may remove finer particles, to leave, eventually, only boulders.

The most devastating mud-flow in recent times happened on 13 November 1985 when an eruption of Ruiz, Colombia, melted ice and snow near the summit. Mud-flows cascaded down the mountain slopes and into the valleys below. One huge flow hit the town of Armero, 50 kilometres from the mountain, killing over 20 000 people, and destroying most of the houses.

31 The town of Armero buried by mud-flows, 1985.

32 Devastation from mud-flows in Armero.

STEAM-ASSISTED ERUPTIONS

Water is very plentiful at and near the Earth's surface, and most volcanoes erupt through water, or through rocks which contain water. The interaction of magma or lava with water or ice, ranging from small steam outbursts to sequences of enormous and devastating explosions, produces very varied eruption effects, and distinctive rock deposits. Virtually all types of eruption (p 10) are at least partly powered by interaction with water. Studies of the 1883 Krakatau explosion (p 56) first suggested that water has a part to play in volcanic eruption.

There is still a lot to learn about ways in which fluids affect magma eruption. Dense carbon dioxide fluid, under pressure, may be a major contributer to certain volcanic explosions. Aquifers – rocks which act as water reservoirs – may contribute to steam explosions near the surface. The flow of magma causes a series of explosions, at ever greater depths. Each explosion releases the pressure of overlying rock, producing a cone-shaped pipe of shattered rock – a tuff pipe, or diatreme – to depths of 2 km or more.

At shallow levels, rapid vaporization of water by magma causes steam to explode with as much as a third of the energy of TNT. Heat may transfer rapidly from the magma to the water in a rapidly expanding and collapsing steam layer at the contact. This breaks up the magma, greatly increasing its surface area, and transfer of heat to the water, until steam expansion overcomes the surrounding pressure and the whole lot explodes. More water mixes with the disrupted magma, feeding a sequence of explosions.

33 Lava-fountains, Réunion Island.

34 Water-assisted explosions at Galunggung, Java.

In fig 33 water has entered the vent to the right. Long streaks reveal the high speed of the water-assisted explosion in contrast with the 'dry' lava burst on the left.

Spectacular steam explosions occur in volcanoes as they build up in shallow seas. Fig 35 shows the way in which steam drives a Surtseyan eruption (p 10), where the vent is just building up above sea-level but sea water spills into the vent. Such activity characterized the emergence of the new volcanic island of Surtsey, off southern Iceland, during the night of 14–15 November 1963. During this type of eruption the vent contains a highly mobile slurry of volcanic, glassy granules mixed with water. As magma erupts through the slurry it mixes with it and rapidly heats the water. Sudden steam expansion shoots the mixture from the vent in sudden jets or in a continuous rocket blast contrasting with quiet, Hawaiian-type lava-fountains (fig 36).

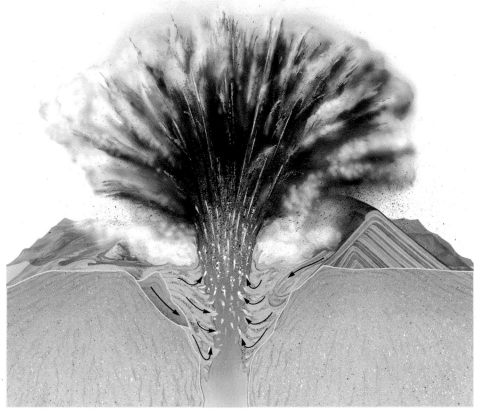

35 Sea water mixing with magma, producing a Surtseyan eruption.

36 Ash, steam and lava at Surtsey.

A VOLCANO IS … A MOUNTAIN?

37 Shishaldin andesite volcano, Alaska, August 1982.

What happens to lava given out during eruptions? Thin runny lava moves away from the crater as long thin lava-flows. Over countless eruptions, the flows build up a flattish mountain shaped like an upturned saucer or circular shield, hence the name shield volcano (fig 39). Lava which is more pasty forms thicker, shorter flows, and builds a steep sided strato-volcano (fig 37). Because pasty lava eruptions are more explosive than runny lava eruptions, strato-volcanoes usually contain a great deal of ash interlayered with the lava-flows. These volcanic cones are prone to collapse as they grow larger. Ash layers are easily eroded by rain, so that a volcano built mostly of ash may be intensely gullied by steep ravines, such as those carved in uncemented ash.

In most people's minds the 'ideal' volcano is a cone-shaped mountain; yet there are many other shapes and types of volcano. Flat regions of plateau lava, widespread ash-flow layers, or lake-filled volcanic depressions equally qualify as 'volcanoes'.

The largest eruptions are caused by the complete foundering of a volcano to form a caldera, an extra-wide crater many kilometres across. Subsidence occurs over a large area during or following an eruption, leaving a huge, shallow depression (fig 38).

The whole of the ocean floor has been built by volcanic eruptions at spreading ridges: the ocean floor may be thought of as an overlapping series of volcanoes, flattened out by the spreading of the floor away from the volcanic centre-line (pp 26–28).

38 A volcano sleeps beneath Lake Taupo, New Zealand.

39 Mauna Kea shield, Hawaii; Mokuoweoweo caldera on Mauna Loa in the foreground.

A VOLCANO IS . . .

Where eruptions repeatedly occur at the same site, the eruption hole is simply a circular crater, often at the summit of a mountain, as at Vesuvius (fig 41) following the 1944 eruption. When eruption ceases, the crater becomes blocked with solidified lava or with rock fragments. Steep interior walls of the crater gradually collapse, and loose rock tumbles into the cavity, so that in time the crater becomes less steep and less deep.

Vesuvius erupted in 1944; half a century later, its crater is still an impressive hole in the ground. Volcanoes which emit runny lava tend to develop a series of pit-like craters with vertical walls, sometimes one inside another (fig 43), or as a chain of craters along a fracture across the volcano.

40 Crater Lake, Oregon, fills a caldera formed after an eruption in about 4000 BC.

41 The crater of Gran Cono, Vesuvius.

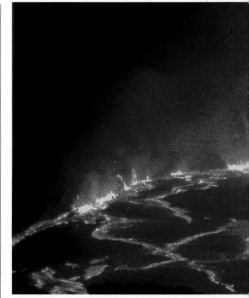

42 Nyiragongo lava lake.

Eruption of lava into a pit crater forms a lake of liquid lava, which crusts over, the crust breaking from time to time to reveal the red-hot interior. Spectacular by night (fig 42), these lava lakes may take many years to cool. Generally speaking, the larger and more explosive an eruption is, the larger the crater it leaves behind. In some eruptions, so much lava is erupted from the magma store below ground that the volcano is no longer supported, and it collapses, creating a caldera. The smallest calderas are about 2 kilometres across. The largest, produced by collapse after eruptions so large it is difficult to imagine them, are tens of kilometres across. The volcanic 'hole in the ground' may be masked by a lake which fills a large depression in the landscape.

43 Pit craters on Karthala, Indian Ocean.

VOLCANOES AROUND THE WORLD

A computer file of volcanoes, published in 1981, shows over 1300 potentially active volcanoes (see p 40). The previous catalogue, compiled between 1950 and 1975, contained a list of about 700 volcanoes. The 'extra' volcanoes were added in the more recent listing for several reasons. First, it was clear that the previous definition of 'active', that is, 'active during historic time', was not useful or workable: 'historic time' was only tens of years in some places, while in others it was a few thousand years. The new list includes all volcanoes known to have erupted in the last 10 000 years.

In the map on this page yet more are added to this total, to account for the innumerable volcanic centres in the ocean depths where ocean crust is generated by volcanic processes at spreading ridges. The map shows a red dot to represent a volcano every 250 kilometres along the spreading ridges in the oceans. It is impossible at present even to guess at how many such volcanic centres are active; we have not seen enough of the ocean floor. The whole ocean floor can be thought of as overlapping, flat, sheet-like volcanoes. These submarine volcanoes, about which so little is known, make up two-thirds of the Earth's surface, the whole of the ocean crust. These volcanoes should be added to the list even though they cannot be seen, and even though they do not make cone-shaped mountains.

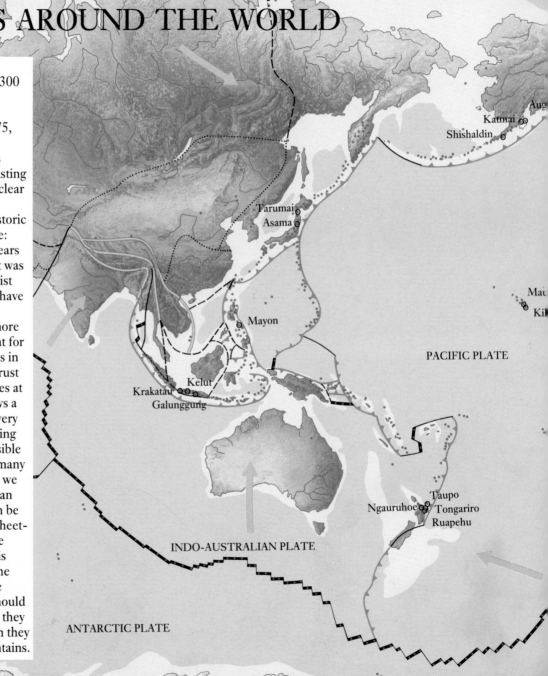

PACIFIC PLATE

INDO-AUSTRALIAN PLATE

ANTARCTIC PLATE

Aug
Katmai
Shishaldin

Tarumai
Asama

Mau
Kil

Mayon

Kelut
Krakatau
Galunggung

Taupo
Ngauruhoe
Tongariro
Ruapehu

NORTH AMERICAN PLATE

EURASIAN PLATE

Krafla
Skaftár
Surtsey

Mt St Helens
Yellowstone

Vesuvius
Lipari Island
Etna
Santorini

Las Canadas

Trou au Natron

El Chichon

AFRICAN PLATE

Mt Pelée

Ruiz

Nyiragongo
Oldoinyo Lengai

Karthala

Chao

Réunion

NAZCA PLATE

SOUTH AMERICAN PLATE

Villarica

ANTARCTIC PLATE

	continental lithosphere
	oceanic lithosphere
	volcanoes
	spreading ridge
	subduction zone
	collision zone with continent
	transform boundaries
	plate boundaries of uncertain type
	other active zones
	movement of plates

27

PLATE TECTONICS, MAGMA

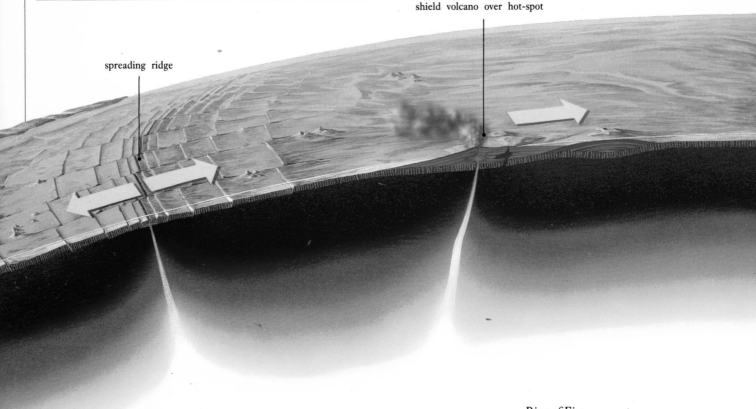

shield volcano over hot-spot

spreading ridge

Spreading Ridges
These are volcanic mountain chains
which form where the Earth's surface is
continually spreading apart. Beneath
ridges, the Earth's solid mantle is rising
towards the surface; pressure release
provides a supply of magma which
separates out and rises along cracks in
the rock. Some magma reaches the
surface, mostly as underwater volcanoes.
While the plates spread apart the magma
cools and becomes part of the plate.

Hot Spots
Hot and active centres within the mantle
produce magma which rises to form
volcano-studded 'hot-spots' at the
surface, both on the ocean floor and on
the continents. Some hot-spot volcanoes
erupt huge volumes of basalt lava. In
places, lines of hot-spot volcanoes are
'strung out' as the plate slides over the
hotter mantle beneath: rising magma
continually 'punches' through the
moving plate.

Ring of Fire
Under the edges of the oceans, the
Earth's plates, including the thin crust,
push down into the body of the Earth.
Their slow descent is marked by
frequent large earthquakes and volcanic
activity. Today this is happening mainly
in the 'Ring of Fire', around the Pacific
Ocean. Magma is made from mantle-
extract and re-melted oceanic crust; it
feeds upwards, melting some of the
continent and becoming more pasty. If
this reaches the surface, gases escape
with difficulty and the magma may foam
up and explode.

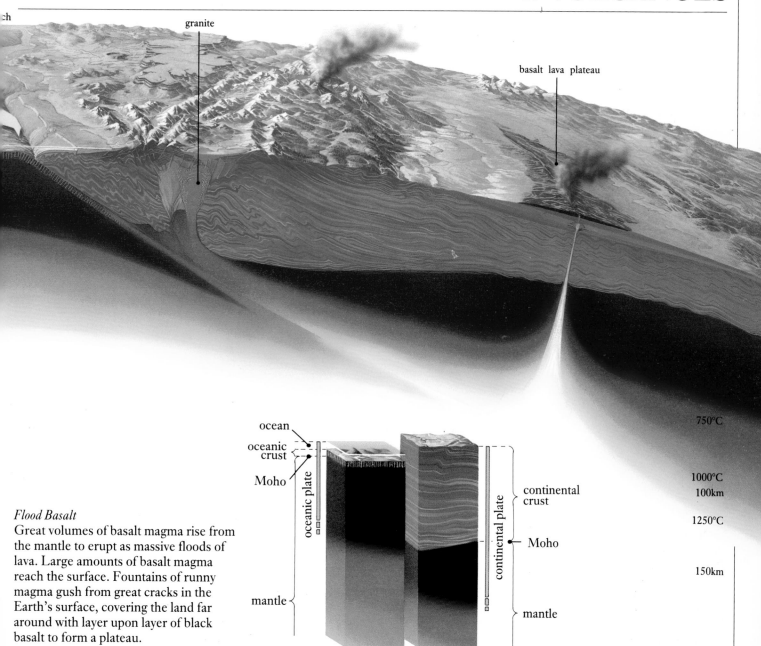

ch

granite

basalt lava plateau

ocean

oceanic crust

Moho

oceanic plate

mantle

continental plate

continental crust

Moho

mantle

750°C

1000°C
100km

1250°C

150km

Flood Basalt
Great volumes of basalt magma rise from the mantle to erupt as massive floods of lava. Large amounts of basalt magma reach the surface. Fountains of runny magma gush from great cracks in the Earth's surface, covering the land far around with layer upon layer of black basalt to form a plateau.

VOLCANOES & PLATES

Spreading ridges are the Earth's greatest volcanic mountain chains, and form where the surface is splitting apart. As circulating mantle rock rises towards the surface, the reduction of pressure in the white-hot rock allows a continual supply of runny magma to separate out and feed along tension cracks. Some magma reaches the surface, mostly at the sea-bed to form underwater volcanoes. However, much of the magma cools deep in the crust, around a long magma reservoir which is constantly stretching apart. As magma continually rises and cools, it becomes part of the plates; newly-formed rock moves away as more magma rises, and the two plates spread apart. As the plates cool and settle, they leave the hot zone as a flat, long ridge, mostly covered by ocean water.

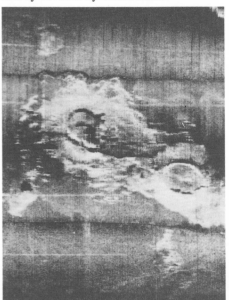

44 Tongariro (foreground), Ngauruhoe and, in the distance, Mt Ruapehu, New Zealand.

45 Ocean floor volcanoes.

Thus, new oceanic crust is formed from magma extracted from mantle rock. The depleted mantle rock makes up the rest of the spreading plate beneath the crust (pp 28–29). The whole plate moves on a weaker, partially-molten mantle zone. Particularly active areas within the mantle create volcanically active 'hot-spots' up to the surface (see p 33), in the oceans (fig 46) or on the continents (fig 47). Some hot-spot volcanoes erupt huge volumes of basalt lava, notably at Hawaii, and at Réunion in the Indian Ocean, where lavas are strung out across the moving ocean floor.

Oceanic plate is eventually 'recycled' back into the Earth's mantle, usually beneath the edges of the oceans. At present most of the Earth's oceanic plate is descending around the edges of the Pacific Ocean. The effects of this include frequent large earthquakes and the violent 'Ring of Fire' volcanic activity from the Andes through Japan and around to New Zealand (fig 44). Where the plate descends, magma from mantle-extract and re-melted oceanic crust feeds into the overlying crust. The rate of magma supply is often enough to melt some overlying crust, altering the magma into a more sticky, silica-rich composition. If this reaches the surface, gases escape with difficulty and the magma may explode (pp 13 and 16) with profound effect upon the Earth's surface, atmosphere and life. Greater volumes of magma, rising rapidly from the mantle, tend to erupt as massive floods of basalt without melting much of the crust on the way up.

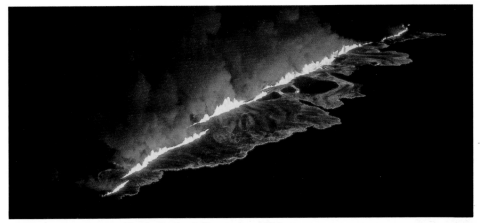

46 Fissure eruption at Krafla, Iceland, September 1987.

47 Trou au Natron, Tibesti Mountains, central Sahara Desert, Chad.

BIRTH OF THE ATLANTIC OCEAN

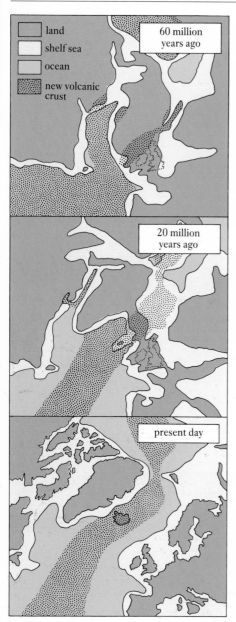

Two hundred million years ago there was no Atlantic Ocean. During the time the dinosaurs roamed the lands, a rift developed in the giant continent of Pangaea, and volcanoes poured out floods of lava. The sea flooded into the valley made by the rift, and a new ocean was born. As volcanic activity continued and the continents drifted apart, more and more ocean crust was created in the gap and the ocean became wider. Some of the lava produced by these volcanoes is on land, but by far the greater amount is under the sea.

The maps (fig 48) show the amounts of lava-covered ocean plate produced since the ocean started to grow. Today Iceland is a unique part of the spreading ridge. It has built up above sea-level as greater amounts of lava are produced at this hot-spot than along other parts of the spreading ridge. In Iceland (p 31, fig 46), there is the opportunity to see how ocean crust is generated. Over historic time, Iceland's volcanoes have revealed something of the nature of volcanoes along the hidden, deep submarine parts of the ocean spreading ridges.

48 Atlantic opening.

49 Lava-flows sixty million years old, Giant's Causeway, County Antrim, Northern Ireland.

A large area of western India is covered with a thick sequence of lava-flows (fig 52), the bulk of which was erupted about 60–65 million years ago. So much lava was erupted in such a short time that activity must have been almost continuous. Some of the flows have an ancient soil layer on them, indicating a period of quiet before the next flow was erupted. Over the last 65 million years, India has moved northwards as the old Tethys Ocean has closed and new ocean plate has been created, covered by the Indian Ocean.

A trail of lavas across this new Indian Ocean floor traces the location of the hot-spot, which stays put as the plate moves over it. The speed of travel of the ocean floor can be discovered from the sequence of ages of the lavas erupted onto it from the hot-spot (fig 51).

The island of Réunion lies above the hot-spot right now. This hot-spot is still one of the most prolific active volcanic sites on Earth. Over the last two million years, Réunion has built into a huge volcanic island, whose summit stands seven kilometres above the sea-bed.

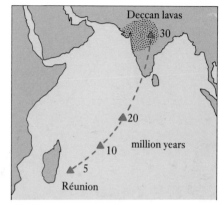

51 Movement over a hot-spot.

50 Rath basalt temple, Ellora, India.

52 Deccan basalt lava-flows, India.

BELOW THE VOLCANO

Within parts of the Earth's mantle, basalt magma gathers and rises through the solid rock. Higher up, within and just below the crust, different kinds of magma collect as reservoirs. Granite is formed where continents collide; basalt rises into a continent, melting some of the crust. Most magma solidifies underground, to form rocks which look very different from erupted volcanic rocks.

Granite and basalt magma produce contrasting volcanoes and eruptions. Above areas where ocean crust descends into the Earth's mantle, huge amounts of granite magma collect in the continental rocks. Red-hot, silica-rich, pasty liquid separates out at the top, sometimes quite near the surface. Granite may be exposed where erosion bites into rising mountains.

Granite magma is sticky and gassy. If it reaches the surface it usually erupts explosively (p 13), particularly if it meets water. Enough may be erupted to weaken the crust overlying the large reservoir; the crust collapses to form a great basin-like caldera (p 25). Hundreds of cubic kilometres of magma can be erupted at once. Present day reservoirs (fig 53) are detected by analysing earthquake waves which have passed through and close by the molten rock. Calderas and their explosion products are recognized from the record in the rocks. Fig 54 shows rocks in the English Lake District which make up enormous blocks of collapsed crust and layers of erupted rock in a Lakeland-wide caldera 450 million years old. Uplift and erosion have enabled us to pry into the complex jumble of rocks.

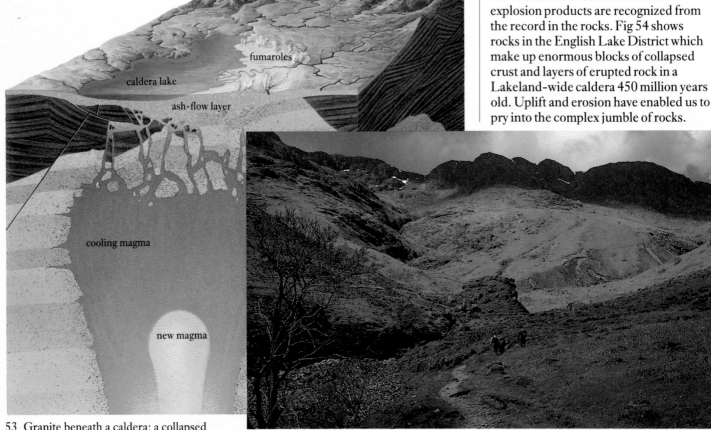

53 Granite beneath a caldera: a collapsed volcano.

54 Inside an ancient caldera: Langdale, English Lake District.

Beneath basalt volcanoes, hot, runny magma is formed where gases and liquid seep up from deep within the mantle. From 60 km below Hawaii, magma collects and slowly rises towards the surface in a tube-like zone (fig 55). Within this zone, narrow 'packets' of magma force upwards through cracks in the rock which re-seal as the magma passes. Near the top of the tube zone, some magma is diverted and forces its way into large cracks across the island volcano. This widens the top of the main pipe like an expanded bottle neck and allows the 'cork' of solid lava to sag into the volcano. The resulting large, steep-walled crater, a summit caldera (p 25), is characteristic of large shield volcanoes (p 22). On Mars there are huge summit calderas. On Earth, magma-feed systems, in the form of infilled cracks, are often seen as dykes and sills exposed by the erosion of ancient volcanic areas.

56 Mauna Loa fissure eruption, Hawaii, 1984.

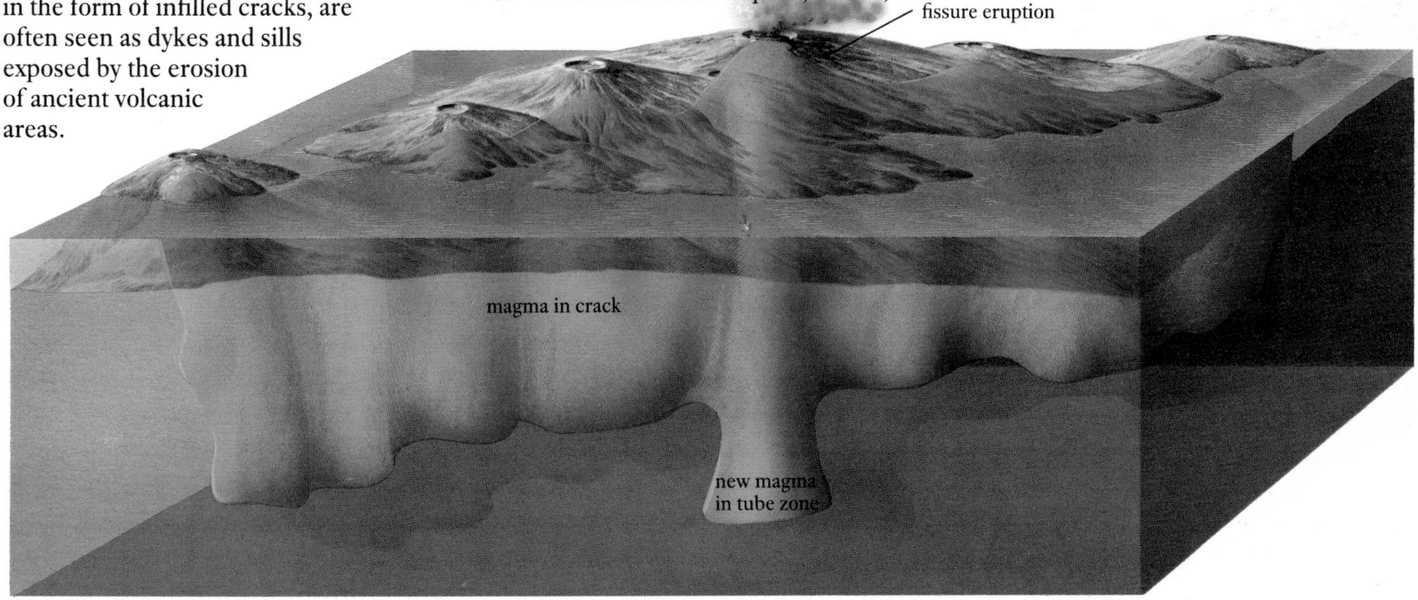

fissure eruption

magma in crack

new magma in tube zone

55 New magma enters a curtain-like crack within a Hawaiian-type volcano.

EDINBURGH VOLCANO

57 Agglomerate (top) and lava.

Central Edinburgh is the site of a group of volcanoes which were active 340 million years ago, in the Carboniferous Period. Repeated eruptions from a number of vents, over a period of a million or more years, spread at least thirteen major lava-flows across the flood plain of a large river. Some volcanic vents, such as the Castle Hill, Crags Vent and Pulpit Rock, only erupted once or twice, and are shown as small cinder cones in the reconstruction opposite. The Lion's Head and Lion's Haunch vents were active for longer, and produced most of the lava now seen in the area. These two vents, only a few hundred metres apart, may have built up a wide but low cone of ash, as in this reconstruction.

The picture (right) shows a sudden, explosive eruption caused by molten lava mixing with water in the vent (see p 20).

Once these volcanoes became extinct, the eroded remnants of their cones were buried beneath river sediment and, later, sea-bed sediments, all of which eventually hardened to rock. Magma was intruded and cooled as flat sheets between these hard rock layers. Where they have been exposed by erosion, the cooled sheets, or sills, now form the Salisbury Crags and the Dassies.

Gradual movements in the crust have tilted the whole volcano to an angle of 25°, while faults sliced the area into blocks. Erosion during the Ice Age left hard rock, the ancient vents, as rugged hills. Softer, ashy layers have been eroded away.

Fig 57 shows a volcanic breccia, made up of broken fragments of basalt lava from the vent of the Lion's Haunch volcano; also a specimen from a lava-flow that erupted from the Haunch vent down the flanks of the volcano to its present resting place on Whinny Hill. It contains cavities where the volcanic gases bubbled out of the molten rock (see p 4).

58 The Lion's Head vent, Arthur's Seat, Edinburgh.

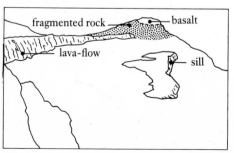

59 Arthur's Seat geology.

60 Edinburgh 325 million years ago and (inset) today.

HOT SPRINGS & GEYSERS

The heat and water in the Earth's rocks together produce a variety of phenomena which include mineral deposits, hot springs and geysers.

Volcanoes, mountains, earthquakes and coastal changes occur as a consequence of the Earth's inner heat and movement (p 28). At the surface, and in the top few kilometres of the Earth's crust, the flow of hot water and dissolved minerals (p 42,44) is a consequence of the Earth's heat and activity in and beneath the crust.

'Geothermal waters' are found mostly in regions of present or recent volcanic activity – frequently around the edges of the Earth's plates, and notably around the Pacific 'Ring of Fire' (p 31). The bulk of geothermal water is surface water stored naturally in permeable rock layers. Some geothermal water is from the surface or sea-bed, or is the water in which the rock was originally deposited. The rocks and water at depth are warmed by the heat inside the Earth.

Geothermal water also comes from molten rock rising towards the Earth's surface. Molten rock carries up heat and dissolved gases; it exudes water, carbon dioxide and other gases as it cools.

A hot spring forms where geothermal water circulates up through permeable rock until it leaks from the ground surface or the sea-bed, while 'smokers' are hot, mineral-rich geothermal springs in the ocean-bed, usually at or near the spreading ridges. Geysers are springs from which columns of boiling water and steam escape into the air from time to time, when dissolved gases foam out, sometimes at regular intervals.

61 Travertine terraces at Minerva Spring, Mammoth, Yellowstone.

62 Mudpot at Midway Geyser Basin, Firehole, Yellowstone National Park.

63 Geyser at Yellowstone National Park, Wyoming.

ACTIVE–EXTINCT–DORMANT?

64 Yellowstone Canyon: fumarole alteration colours rocks.

Is a volcano dead or alive? Is it safe to live and farm on it? Unfortunately we live in a time scale which is so short that it is difficult for us to assess volcanoes well enough to be sure when they are truly dead and 'safe'. Published lists of volcanic activity may take a date – such as 10 000 years ago – and state that if a volcano has not erupted since that time it is extinct. Yet the site of a large explosive eruption can be a lake-filled area where eruptions happen every few thousand years, or tens of thousands of years, from various parts of a caldera tens of kilometres across. Such eruptions are extremely unpredictable. Intervals of quiet are so long in relation to our period of observation that it is impossible to tell whether the area has undergone its very last great explosion.

65 Cinder cones and lava-flow, Flagstaff, Arizona.

At the other extreme, a volcano may erupt only once, to form a cone which is eventually eroded or covered over. Some rock layers, millions of years old, contain the remains of old volcanoes (p 36); these are certainly extinct. However, as long as there is nearby volcanic activity or geysers and hot springs, or if there have been eruptions within the past few thousand years, it remains uncertain whether there will be another eruption within the next few thousand years – or even within our lifetimes. Whether a volcano might be thought of as 'dormant' depends upon the likelihood of magma again rising to the surface. This movement can be monitored (p 50); but in most areas there is no long term information: civilization and technology are too recent for us to know all of this planet well enough.

67 Sulphur fumarole, New Zealand.

66 The crater of Oldoinyo Lengai, Tanzania.

HOT METAL, MINERALS & WATER

Much of the rock in the Earth's crust contains water. Water can move through rock, dissolving minerals and carrying them along.

Reservoirs of hot or molten rock warm the surrounding rock and the water in it; this warmed water rises through the solid rock, seeping through cracks and pores.

Direct contact of magma with surrounding rocks, particularly limestone, creates zones where the magma reacts chemically with the rock (a). 'Contact mineral deposits' are formed, such as those of tungsten minerals (pictured, lower right). Magma contains water, carbon dioxide and many other volatile substances. Some of these are expelled as fluid mixtures while the molten rock slowly cools. Fluid from within the magma can pervade and alter the surrounding rocks or be injected as 'pegmatite dykes' (b), depositing minerals such as topaz and beryl. Sometimes the outer, solidified parts of a cooling magma reservoir form a hard 'shell' around the still-liquid centre, sealing in the volatile gases exuding from the solidifying rock. The pressure inside rises until the shell and surrounding rocks shatter in a huge, exploding wave of small fractures (c). The released fluids –

68 Hot rock, hot water and minerals.

often including molybdenum and copper – deposit their mineral load in the fractures (pictured, lower left) and greatly alter the rock. Fig 69 shows a huge mine in a copper deposit which formed in this way.

Minerals may be deposited from warm water circulating in the crust, even if the minerals are carried in the water at very low concentrations. Over many thousands of years, considerable mineral deposits build up (d), such as those of galena (pictured, top right) and quartz. Some water travels through cooled magma, collecting further minerals and metals such as silver and gold (pictured, top left) from the igneous rock and redepositing them elsewhere (e).

Magma reaching the surface releases much of its load of volatile substances, mainly in the form of gases such as steam, carbon dioxide, hydrogen chloride and sulphur dioxide. Lava and other volcanic deposits contain a solid or liquid residue of these and, together with metals and other substances, they may become concentrated. Some may be transported by rainwater and redeposited as minerals, either within the volcanic rocks or in new sediment layers elsewhere (f), notably in boron- and soda-rich lakes.

Springs and geysers issue where circulating water reaches the surface or the sea-bed. Minerals such as zinc blende and pyrite are deposited from the spring water within the soft, cooler sediment layers (g).

69 Chuquicamata opencast copper mine, Chile.

USEFUL HOT WATER

The Earth's natural heat is gradually being lost. Although this planet is mostly white-hot inside, heat flowing up through the surface has only a very slight effect upon the atmosphere. However, as any miner knows, the Earth's inner heat is quite obvious when you descend a little way into the crust. The temperature exceeds one hundred degrees Celsius only ten kilometres down. 'Geothermal waters' (p 38) are found in regions of present or recent volcanic activity, mostly around the edges of the Earth's plates, where heat flow is greater.

Most of the Earth's heat flow is too diffuse for use as an economic alternative to non-renewable fuels. Small areas have enough useable hot water to be converted for electric power or district heating systems. In order to be of economic use, this heat has to be concentrated in thermal reservoirs where it has been built up and stored by natural processes. Naturally-hot water has been used since ancient times. Hot springs have been the sites of medical, bathing and recreation centres – as well as kitchens – for thousands of years.

'Heat fields' and 'heat wells', rather than oil fields and wells, could provide more energy as oil becomes more and more scarce and its cost rises. In the same way that oil fields are explored, drilled and developed, geothermal fields will be surveyed, drilled and discovered, assessed (for corrosive gases, heat output, etc.) and developed to extract heat energy. Prospects include areas of recently active volcanoes; those in or near active areas and calderas (p 25); solid granite masses and 'wet' sediments at depth.

70 Geothermal plant at Wairakei, New Zealand.

In some areas, dense, super-hot, watery fluid at depth is extracted as steam to drive turbines for electricity production. Elsewhere, hot water is pumped to provide district or local heating for domestic or agricultural uses. A few geothermal areas are well-established; others are only now being evaluated. Important sites include Iceland, Italy, New Zealand, Japan, eastern USSR and western USA. Geothermal systems are depleted by usage. An extraction well may be artificially stimulated, for example by shattering rock deep inside a borehole with explosives, to make it more permeable.

Areas underlain by suitable volumes of hot, but dry, rock may be drilled to set up a water pumping system to extract the geothermal heat energy from the rock.

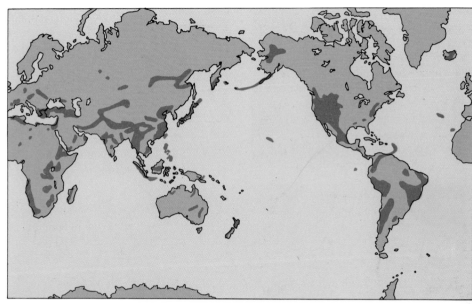

72 Geothermal regions of the world.

71 Geothermal bathing in Japan.

73 Growing flowers and tropical fruits in Iceland.

USEFUL ROCKS

74 Ignimbrite building stone.

76 High altitude sulphur mine, Bolivia.

Volcanoes have a bad press: in the media they are killers, destroyers of homes and land, so it is easy to overlook their usefulness. Volcanic areas may have highly fertile soils, supporting lush agriculture (fig 75). While a deluge of volcanic ash spells disaster for crops, a light dusting may add valuable chemicals to soil. Quarried and marketed as 'soil improver', ash can add nutrients and upgrade the texture of some soils. Pumice is used as industrial abrasive. The white cliffs of pumice at Lipari (fig 80) have been worked for centuries. Volcanic ash, being full of cavities, is useful as a lightweight aggregate, and was exploited as such by the Romans. The Romans used pozzuolana ash, which sets under water, as hydraulic cement.

75 Lush agriculture on volcanic soil, Canary Islands.

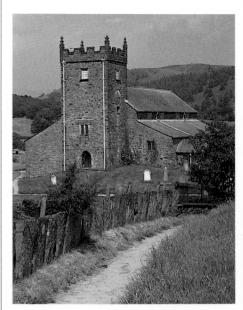
77 Lakeland volcanic building stone.

Beds of volcanic ash, hardened over time, can be carved into blocks for use as lightweight building materials. At the same time, the method of working the stone provides a cavity in the ground which, covered with a façade, is suitable as a cool, well-drained dwelling (fig 78). Sometimes the volcanic ash may be decoratively carved (fig 79). Basalt and other lavas are used as construction material, varied colours being exploited decoratively by the builders (fig 81). The best known volcanic material used for building in the United Kingdom is the 450 million year-old volcanic ash from the Lake District (fig 77 and p 34). Useful minerals (p 42) are associated with volcanoes. The world's highest altitude mines, in the Bolivian Andes, work volcanic sulphur (fig 76).

79 Church gateway, Arequipa, Peru

81 Lava and pale ash, Santorini.

78 Cool homes in volcanic ash.

80 Pumice-works at Ponte Pelato, Lipari island.

VOLCANO HAZARD

A prediction of a volcanic eruption must first of all say what is likely to happen: local decision-makers and residents need to know about the hazard they face. Those living nearby need to know if their daily lives will merely be interrupted and inconvenienced, or if their lives are at risk and whether it is possible that their property might be destroyed. Will the eruption be a small event confined to the flanks of the volcano or will it be violent, devastating a large area? Fig 83 shows how different types of volcanic activity present a hazard over different areas. Lava-flows are a hazard to property, but are usually confined to the slopes of the volcano itself. At volcanoes which are frequently active, lava-flows are usually well understood by local residents. The most widely distributed hazard is ash-fall, which may extend a thousand kilometres or more from the volcano.

At its worst, ash brings total darkness for hours, suffocating animals, smothering plant life and preventing the use of machinery.

Extending tens or hundreds of kilometres around the volcano, pyroclastic-flows and mud-flows are the greatest hazard (see pp 14 and 19). Developing rapidly, and capable of travelling at speeds up to 200 kilometres an hour, these flows are responsible for almost all deaths from volcanic eruptions. They can wipe out towns in a matter of minutes. Understanding pyroclastic- and mud-flows is of paramount importance in reducing deaths. An idea of the distance over which mud-flows cause destruction is seen in fig 82 where a lava flow from the volcano Villarica, visible in the distance, is advancing over an area previously scoured by mud-flows which wiped out a saw mill.

Volcano catalogues show what seems to be an alarming increase in the number of volcanoes and their activity. Volcanologists believe, however, that global volcanic activity is probably neither increasing nor decreasing over

82 Lava-flow over a mud-flow from Villarrica, Chile, 1971.

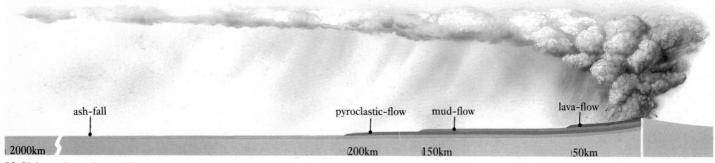

ash-fall   pyroclastic-flow   mud-flow   lava-flow

2000km   200km   150km   50km

83 Volcano hazards at different distances from an eruption.

historical time. What the records are showing is an increase of knowledge of planet Earth and its processes. Rapid increases in the number of volcanoes recorded represent times of exploration and discovery – the first being between 1500 and 1550 (fig 86) when Europeans were discovering the Americas. The painting of lava-flows threatening the Sicilian town of Catania (fig 85) dates from this time. Travellers would have recognized the same lava hazard at 'new' volcanoes in the New World. The second period of rapid increase coincides with another great time of travel and scholarship in the eighteenth and nineteenth centuries. Archaeological discoveries triggered interest. Hordes of tourists witnessed eruptions of Vesuvius, which so many centuries before had wiped out the town of Pompeii (fig 84). The number of eruptions reported diminishes during times of social disaster, such as war or famine. Media coverage of eruptions increases after a large volcanic disaster, such as Krakatau, Mont Pelée, or Mt St Helens. Modern communications, by their immediacy, tend to emphasize eruptions in developed areas.

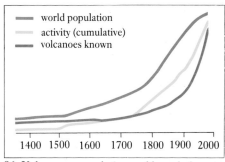

86 Volcanoes, population and knowledge.

84 Isis Temple, Pompeii, uncovered 1765.

85 Etna, 1669; lava-flows from Monti Rossi threaten Catania.

VOLCANO WATCHING

87 Observing Mount St Helens.

Volcanologists try to forecast a volcano's likely activity by looking at its past activity – assuming that it will continue to erupt as before. The record includes eyewitness accounts, as well as the rock record left by these and earlier eruptions. This can be supplemented by instrumental recordings of current eruptions. For example, tiltmeters record changes in a volcano's slopes; the rise of magma into the volcano causes it to swell; cessation of the eruption is heralded by subsidence (fig 89). After 60 years of observations at the Hawaiian Volcano Observatory it is possible to make accurate forecasts of activity.

Hawaiian eruptions are frequent and usually quite mild. By contrast, volcanoes like Mount St Helens erupt infrequently but violently. To minimize loss of life it is important to identify volcanoes which are coming to the end of a long period of quiet, which is difficult, and then to monitor them intensively, which is expensive.

The 1980 eruption of Mount St Helens was small compared with past eruptions, evidence of which is preserved in the geological record (see pp 14 and 59). It was correctly identified as reaching the end of its quiet period. The eruptive cycle that reached its climax on 18 May 1980 was recorded and photographed from the onset of premonitory signs (p. 59), through to the last stages of activity. The results will help in understanding future eruptions of this dangerous type.

88 Volcano observatory near Halemaumau crater, Hawaii.

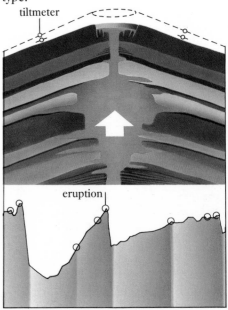

89 Tiltmeter monitoring record.

When volcanic hazard is assessed, and an eruption is imminent, what can be done to minimize the effects of the eruption? Lava-flows are likely to follow existing valleys, though flows may be diverted by building barriers, or slowed with cooling water (fig 91).

Explosive eruptions pose hazards which can neither be contained nor directed; if an eruption is likely, hazardous areas need to be identified and evacuation considered, despite the social disruption. Plans should include adequate morgue facilities, and local hospital emergency facilities for victims with burns or with lungs damaged by hot ash. Important too are a supply of face masks for the general population and breathing apparatus for emergency workers, with advice to local communities on how to cope if cut off for several days.

Following the Mount St Helens eruption of 1980, there are local guidelines for coping with heavy ashfall. Residents should be advised of these, including the importance of sweeping ash from roofs to prevent collapse. Authorities should be equipped to measure levels of toxic gases and analyse ash particle size. Alternative sources of drinking water should be located. Roads and railways may be cut by mud-flows or lava-flows; aircraft and cars may be unable to operate for days because of airborne ash; motors of all kinds may be unusable. Ashfall may also interfere with radio and TV transmission.

It is not surprising that after a major explosive eruption it is very difficult to find out what is going on and arrange help.

90 Shopping in Washington State, May 1980.

92 Sign on bank door, Oregon, 1980.

91 Heimaey's harbour shape modified by water-cooled lava.

TAUPO AD186

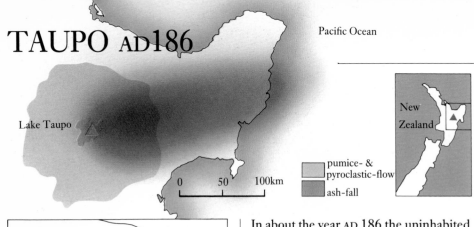

Pacific Ocean

Lake Taupo

0   50   100km

New Zealand

pumice- & pyroclastic-flow

ash-fall

1800 years ago

3400

5400 — 22500

rubble

road

93 Ages of pumice layers in fig 94.

In about the year AD 186 the uninhabited lands of New Zealand were shattered by the largest explosion the world has known for thousands of years. Lake Taupo (p 23, fig 38) hides the site of this volcanic outburst: it is one enormous, active volcano in a bowl-like depression.

The most recent in a long history of explosions in the Taupo area, the eruption started from beneath the lake waters with a long blast of gases and ash.

As the vent became free of water, there followed a larger, Plinian eruption with an ash cloud about 30 km high. Lake water mixed with the erupting column of hot ash and pumice; thick layers of ash spread across the land before the eruption was quenched.

However, the worst was yet to come. After a quieter spell, when the lake was dried out or expelled by smaller eruptions, more ash was spread around the volcano. The large vent, now free of water, continued to erupt with ever-increasing energy until a cloud of ash rose 50 km in an exceptionally violent eruption of more than 20 cubic km of pumice. Part of the eruption column collapsed. It spread across the land as a pyroclastic-flow (p 14). With such massive ejection of magma, the whole vent area collapsed into the top of the magma reservoir – and set off an even more catastrophic explosion. Thirty cubic km of hot pumice, ash and rock fragments were erupted outwards at over 300 metres per second, flowing like liquid over 1500 metre mountains, to cover an area 160 km across in just seven minutes. Clouds of ash covered 30 000 square kilometres.

The eruption was over. Within 20 years Taupo Lake had refilled, while the area continued to settle during severe earthquakes. The course of this eruption is revealed by detailed studies of the rock layers (fig 94); these show that it is not an unusual catastrophe in the immensity of geological time. However, it is obvious from the evidence of past explosions that neither the timing nor the size of the next eruption can be predicted.

94 Road cutting in Taupo ash.

SANTORINI *c.* 1645BC

Viewed from the air (right), the group of islands called Santorini, or Thera, form a broken circle; they are the remnants of a larger island which was fragmented during a huge volcanic eruption over 3500 years ago. A volcano had been in existence here for a million years, and a long period of dormancy, about 15 000 years, preceded this massive eruption. From the ash layers, we can tell the eruption began with a Plinian explosion; an eruption column about 30 km high spread ash over the eastern Mediterranean, and buried the immediate area of the island under several metres of pumice blocks and ash.

In a locality in eastern Turkey, 24 cm of ash have been identified, so the map may show an underestimate of the ash-fall area. The Plinian explosion widened the crater of the volcano, until, eventually, sea water entered the crater, triggering more violent explosive activity, and greater fragmentation of the lava.

The eruption has been dated by various means; from excavations of the city of Akrotiri which was overwhelmed by the eruption, twigs and seed have been dated using radiocarbon; and further afield, a layer of acid ice preserved in the Greenland ice cap has given the same age. The world's weather must have been dramatically affected. Such a large eruption, once dated, provides a widespread time marker for both archaeologists and geologists.

The volcanic islands at the centre of the bay may in time build up into a new, large volcano which may explode in the same way.

 ash-fall

95 Caldera formed after the 1645 BC eruption of Santorini.

VESUVIUS AD79

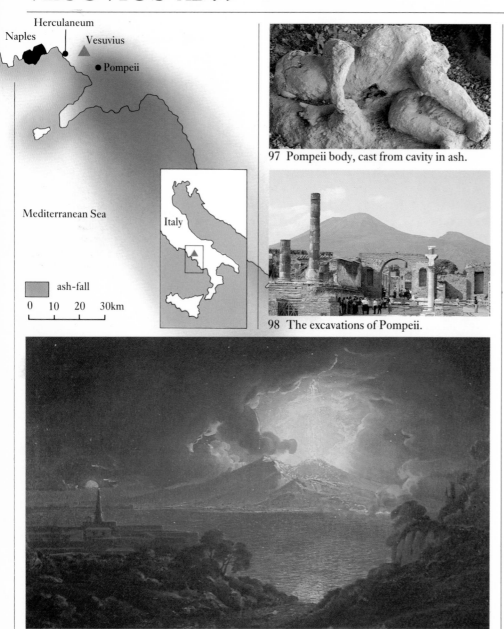

Herculaneum
Naples
Vesuvius
Pompeii

Mediterranean Sea

Italy

ash-fall

0   10   20   30km

97 Pompeii body, cast from cavity in ash.

98 The excavations of Pompeii.

Until its violent eruption in August, AD 79 it was not realised that Vesuvius was a volcano. The huge cloud which rose from the mountain was, for the Romans, an event without precedent. The eruption is recorded by Pliny the Younger in a letter to Tacitus, the earliest known eye-witness account of any eruption. He tells how the wealthy towns surrounding the volcano were darkened by falling ash and later engulfed in pyroclastic-flows and mud-flows. Excavations revealed body-shaped cavities and skeletons of about 2000 people in Pompeii alone (fig 97), though the majority of the population fled. Causes of death are interpreted from casts of human forms. Some showed fractured skulls, others were overwhelmed by ash or noxious gases. Some survived falling ash, only to be overcome by pyroclastic-flows. Eighteenth century digging chanced upon treasure from the wealthy Roman towns. Excavations began, amid intrigue and the smuggled export of art objects. The Temple of Isis (p 49 fig 84) contained the bones of an animal whose sacrifice had been interrupted by the eruption. In the nineteenth century, systematic excavation revealed the lost civilization as well as the events of the eruption.

During the eruption, the old mountain was largely destroyed. A new cone has built up, encircled by the remnants of the old cone. The last eruption was in 1944; since then, Vesuvius has been quiet. It is not known whether it will remain dormant for tens, hundreds or even a thousand years.

96 Eruption of Vesuvius, 1774, from a painting by Joseph Wright of Derby.

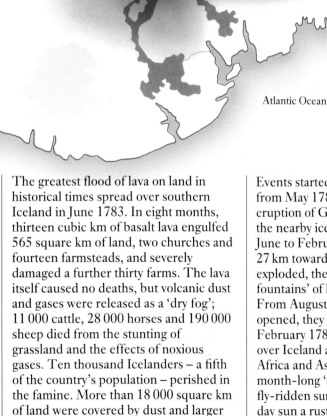

ice-cap
ash-fall
lava-flow
fissures

0   10   20   30km

Grímsvötn

Laki

Iceland

Atlantic Ocean

99 The Skaftár fissure.

The greatest flood of lava on land in historical times spread over southern Iceland in June 1783. In eight months, thirteen cubic km of basalt lava engulfed 565 square km of land, two churches and fourteen farmsteads, and severely damaged a further thirty farms. The lava itself caused no deaths, but volcanic dust and gases were released as a 'dry fog'; 11 000 cattle, 28 000 horses and 190 000 sheep died from the stunting of grassland and the effects of noxious gases. Ten thousand Icelanders – a fifth of the country's population – perished in the famine. More than 18 000 square km of land were covered by dust and larger fragments. An eyewitness record of the eruption was made by Revd Jon Steingrimsson from Prestbakki, a village almost surrounded by the lava.

Events started with earthquake activity from May 1783 and possibly with an eruption of Grímsvötn volcano beneath the nearby icecap, Vatnajökull. From June to February, ten fissures opened for 27 km towards Grímsvötn. Each exploded, then erupted huge 'fire-fountains' of lava which filled valleys. From August, when the last fissure opened, they all erupted together until February 1784. A bluish haze settled over Iceland and spread to Europe, Africa and Asia. In southern England a month-long 'smoky fog' gave a stifling, fly-ridden summer, turning the noon-day sun a rust red colour. Following winters were very severe in Europe. The Skaftár fissure landscape (fig 99) is marked by over a hundred craters, with Laki forming the high-spot.

KRAKATAU 1883

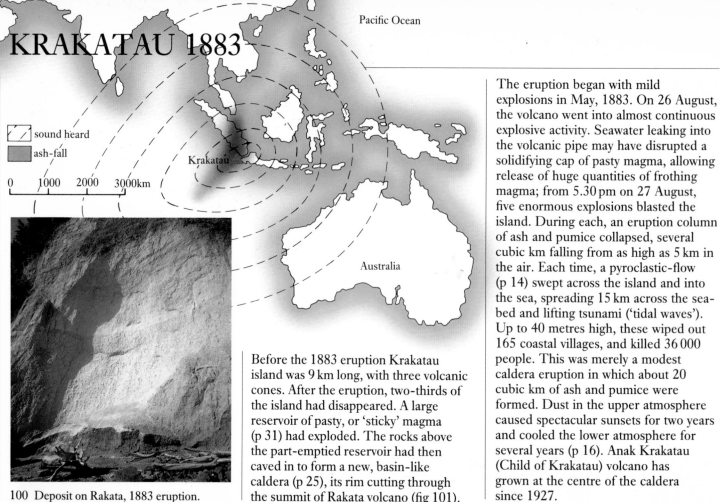

Pacific Ocean

sound heard

ash-fall

0   1000   2000   3000km

Krakatau

Australia

100 Deposit on Rakata, 1883 eruption.

Before the 1883 eruption Krakatau island was 9 km long, with three volcanic cones. After the eruption, two-thirds of the island had disappeared. A large reservoir of pasty, or 'sticky' magma (p 31) had exploded. The rocks above the part-emptied reservoir had then caved in to form a new, basin-like caldera (p 25), its rim cutting through the summit of Rakata volcano (fig 101).

The eruption began with mild explosions in May, 1883. On 26 August, the volcano went into almost continuous explosive activity. Seawater leaking into the volcanic pipe may have disrupted a solidifying cap of pasty magma, allowing release of huge quantities of frothing magma; from 5.30 pm on 27 August, five enormous explosions blasted the island. During each, an eruption column of ash and pumice collapsed, several cubic km falling from as high as 5 km in the air. Each time, a pyroclastic-flow (p 14) swept across the island and into the sea, spreading 15 km across the sea-bed and lifting tsunami ('tidal waves'). Up to 40 metres high, these wiped out 165 coastal villages, and killed 36 000 people. This was merely a modest caldera eruption in which about 20 cubic km of ash and pumice were formed. Dust in the upper atmosphere caused spectacular sunsets for two years and cooled the lower atmosphere for several years (p 16). Anak Krakatau (Child of Krakatau) volcano has grown at the centre of the caldera since 1927.

101 Island of Rakata, remnant of the island of Krakatau.

VALLEY OF TEN THOUSAND SMOKES 1912

102 Ravine eroded in the 1912 deposit.

103 Level plain produced by the 1912 pyroclastic-flow.

This eruption took place in June 1912, and is one of the most violent ever recorded. The volcano Katmai is in a sparsely inhabited region, and there were no records or even legends of it being active. Earthquakes on 2 June were felt at Katmai village, 15 km away, and again on June 4 and 5, but as earthquakes are common in this part of the world, they caused no alarm.

The first major explosion took place at 1 pm on 6 June, and was heard 1500 km away. Further large explosions took place later that day and the following morning. A dense ash cloud enveloped much of Alaska, producing total darkness which at Kodiak lasted 60 hours. Acid gases released in the explosion burned the throats and lips of inhabitants, making eating impossible; it even damaged washing hanging out to dry in Vancouver 2500 km away. The sea became clogged with floating pumice, so thick that it could support the weight of people walking over the water! Fine ash was carried far from the volcano, producing a reddish haze in Algiers. National Geographic Society expeditions found the Valley (fig 103) carpeted with new pyroclastic-flow deposits to a depth of 50 metres (fig 102). Innumerable fumaroles, the inspiration of the name of the Valley, rose to heights of several hundred metres; over the years their activity has declined.

This eruption was a landmark in our understanding of pyroclastic-flow eruptions: here the connection was made between a collapsed volcano and the catastrophic eruption of fragmented lava.

Mt Katmai

Pacific Ocean

Alaska

Anchorage

ash-fall

0   50   100 150km

MONT PELÉE 1902

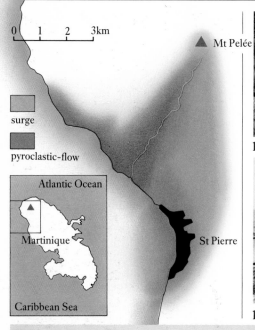

surge

pyroclastic-flow

Atlantic Ocean

Martinique

Caribbean Sea

Mt Pelée

St Pierre

105 Ciparis' dungeon.

106 St Pierre, 1902; Mont Pelée spine.

104 St Pierre and Mont Pelée in 1987.

After two years of increasing fumarole activity in the summit crater, Mont Pelée erupted in late April, 1902. Several water-assisted explosions erupted black ash and steam, accompanied by mud-flows and floods, until 7 May. By the end of 6 May, a pasty magma plug had risen to the base of the crater: a glow, with red-hot blocks, was seen from St Pierre town. Small pyroclastic-flows (p 14) descended the south-west flank. Then, at 8 am on 8 May, a fast, dark, ground-hugging flow blasted St Pierre town, killed 28 000 inhabitants within seconds, sweeping on across the harbour and setting fire to ships. This flow of gas, ash and pumice had burst with explosive force from beneath the growing magma plug within the volcano and had been directed through a notch in the crater. The flow's momentum and gassy turbulence carried it at great speed, to burst upon the town. Burning hot and choking, it filled every room and cavity as the gases expanded into them. The only survivor in the zone of total destruction was Louis-Auguste Ciparis, a prisoner protected in a dungeon (fig 105); he was badly burned. A further flow completed the destruction of St Pierre on 20 May. Later eruptions culminated in a flow on 30 August which doubled the area of devastation, after which eruptions decreased in size and frequency into 1904. Ash had spread across the whole of Martinique. The lava dome grew for a year, with a conspicuous lava 'spine' (fig 106). In 1902 the dangers of this volcano were greatly underestimated; the most recent previous eruption had been before the first European settlement of 1635.

MOUNT ST HELENS 1980

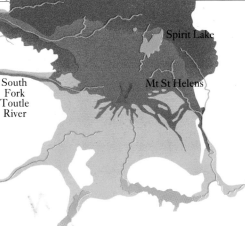

North Fork Toutle River

South Fork Toutle River

Spirit Lake

Mt St Helens

devastated area

predicted hazard area

overlap

0   10km

North America

Pacific Ocean

In 1978, the United States Geological Survey reported that Mount St Helens might erupt violently before the end of this century, affecting human life and health, property and agriculture over a broad area. It defined expected types of activity and hazard zones, advising a monitoring programme. The eruption, one of the most closely-observed and well-documented ever, happened in 1980. In March, earthquakes were followed by small explosions. The volcano began to bulge; it was clear that magma was rising. On 18 May at 8.23 am, the bulging slope could no longer contain the magma; the entire north flank of the volcano quivered, appeared to liquefy, and burst apart in a landslide that released the climactic eruption. A gas blast from exposed magma broke through, overtook the landslide, and in minutes devastated an area 30 km wide and extending 20 km northward. It stripped an inner zone of trees, and flattened surrounding forest (fig 107). The blast cloud became airborne, leaving a narrow zone where trees were seared or singed. A continuous explosion lasting 9 hours threw ash 25 km into the atmosphere. The ash cloud moved eastwards, causing darkness, some ash settling out in the Great Plains 1500 km away. Devastating mud-flows along river valleys were fed by water from melted glacier ice and snow, and from Spirit Lake. In the new, enlarged crater, extrusion of lava domes alternated with explosions and pyroclastic-flows which decreased in intensity into 1986; it may be years before the volcano becomes quiet.

107 Tree 20 km from the crater.

108 Mount St Helens, July 22 1980.

INDEX

aa, 5
activity levels, 26
Akrotiri excavations, 53
andesite, 6
ash, 6, 13, 22, 46, 53, 57
atmospheric effects, 16, 55, 57

basalt, 4, 28, 29, 31, 35
blocks, 13
bombs, 13
building stones, 47

caldera, 22, 25, 34, 35
carbon dioxide, 8, 20, 43
cave, 5
chlorine, 8
climatic impact, 16–17, 55, 56
cone, 10, 13, 22
copper deposits, 43
crater, 14, 24, 25
crumble breccia, 6
crystals, 2

dacite, 6
damage caused by lava, 5, 6, 13, 19, 46, 48, 54–9
Deccan Plateau, 5, 33
destruction caused by lava see damage
diatreme, 20
distance of flows, 5, 6, 52
dome, 6
dust-veils, 16
dyke, 35

earthquakes, 28, 57

fire fountain, 4, 13
flood basalt, 29
fluorine, 8

galena, 43
gases, 1, 2, 8, 43, 55, 57
geothermal energy, 44–5
geothermal waters, 38, 44
geysers, 38
glass, 18
gold, 43
granite, 34

Hawaiian type activity, 10
hazards, 48–9, 51
heat energy, 44–5
hot spot, 28, 31, 32, 33
hot springs, 38, 44
hyaloclastite, 18

hydrogen chloride, 43
hydrogen sulphide, 8, 16

Ice Age, 36
ignimbrite, 14

lahar, 19
lapilli, 13
lava, 1
lava lake, 25
lava-fountain, 21

magma, 1, 28, 30, 31
mantle, 28, 30, 31
Mars, 35
mineral deposits, 38, 42–3
mud-flow, 19

obsidian, 6
ocean floor eruptions, 4, 18, 22, 26, 28, 30

pahoehoe, 5
Pangaea, 32
pegmatite, 42
Pelés hair, 13
Phreatoplinian type activity, 10
pillows, 5, 18
plate movements, 28, 30, 33
plateau, 14, 22, 29
Plinian type activity, 11
Pompeii excavations, 49, 54
pozzuolana ash, 46
prediction of eruptions, 40–1, 48, 50
pumice, 13, 46, 52
pyrite, 43
pyroclastic-flows, 6, 14, 48, 52, 57
pyroclastics, 13

rhyolite, 6
Ring of Fire, 28, 31

seawater action on lava, 18, 21, 53, 56
shield volcano, 22
silica, 6
sill, 35, 36
silver, 43
smokers, 38
soils on lavas, 33, 46
speed of flows, 5, 6, 10, 13, 14, 52
spine, 6
spreading ridge, 22, 26, 28, 30, 32
steam, 8, 18, 43
strato-volcano, 22
Strombolian type activity, 10
Sub-Plinian type activity, 10

sulphur, 8, 16, 47
sulphur dioxide, 16, 43
sulphuric acid, 16
Surtseyan type activity, 10

temperature, 2, 8
tephra, 13
Tethys, 33
Thera, 53
thickness of flows, 5, 6
trachyte, 6
tsunami, 56
tuff pipe 20

Ultraplinian type activity, 11

viscosity, 2, 6
volume of flows, 14, 52, 56, 57
Vulcanian type activity, 11

water action on lava, 18, 21, 53, 56
weather changes, 16–17, 55, 56

zinc blende, 43

Designer: David Robinson
Artists: Gary Hincks, Mike Eaton
Typesetting by Cambridge Photosetting Services
Printing & Binding in Singapore by Craft Print Pte Ltd

10 9 8 7 6 5 4 3 2 1

Picture sources:
Front cover, inside front cover, Figs 3, 4, 5, 6, 8, 9, 10, 15, 16, 23, 24, 25, 26, 28, 29, 30, 31, 33, 34, 37, 39, 40, 41, 42, 44, 56, 61, 62, 63, 64, 67, 70, 71, 87, 88, 98, 99, 100, 101, 102, 103, 107, 108 Katia & Maurice Krafft, BP 5, 68760 Cernay, France
Figs 11, 19, 20, 21, 22, 27 Stephen Self, University of Texas Arlington
Figs 12, 13, 36 John E. Guest, University College London
Fig 14 J. G. Moore, United States Geological Survey
Fig 17, 76 Peter J. Francis, Open University
Fig 18 T. Miller, United States Geological Survey
Fig 32, Colorific
Fig 38, 78, 94, 104, 105 Colin Wilson, Bristol University
Fig 43 Brian Upton, Edinburgh University
Fig 45 Institute of Oceanographic Sciences, Godalming, Surrey

Fig 46 Eysteinn Tryggvason, University of Iceland
Fig 47 Pesce
Fig 49, 58 BGS Keyworth, Nottinghamshire
Fig 50 D. Traverso
Fig 52 Keith Cox, Oxford University
Fig 54 Ian Mercer
Fig 57 84, 85, 106, NHM
Fig 90, 92 Sonia Buist
Fig 60 Aerofilms
Fig 65 Peter J. Adams
Fig 66 Barry Dawson, Edinburgh University
Fig 69 Moyra Gardeweg, Servicio Nacional Geologia e Mineria, Chile
Fig 73, 91, inside back cover, F. K. Frhr. v. Linden, Waldsee, Germany
Fig 74, 80 H-U Schmincke, Kiel University
Fig 75, 97 Picturepoint
Fig 77 Eric Robinson, University College London
Fig. 79 George Walker, University of Hawaii
Fig 81 Diana Smith
Fig 82 James Stewart
Fig 95 Tim Druitt, University College of Wales Cardiff
Fig 96 Private Collection, UK

The authors wish to thank many volcanologists and geologists for their help and advice in the preparation of the text and artwork, in particular Janine Baldock, Alistair Baxter, Peter Baxter, Mike Branney, Joe Cann, Angus Duncan, ' Guest, Chris Kilburn, Peter Kokela Stephen Self, John Thackray, Colin Wilson, Christine Woodward, Brian Ur

Library of Congress Cataloging Publication Data
Van Rose. Susanna.
   Volcanoes / by Susanna van Rose and Ian Mercer. – 2nd. ed.
      p. cm.
   Originally published in 1974 by H.M.S.O. for the Institute of Geological Sciences, London.
   Includes Index.
   ISBN 0-674-94307-4
   1. Volcanoes   I. Mercer, Ia
   II. Title.
QE521.2.V36   1991
551.2'1 – dc20